THIS BOOK BELONGS TO

TABLE OF CONTENTS

I see 1 starfish.

1 1 1 1 1 1

One One

1 1 1

PRACTICE HERE

PRACTICE HERE

I

I see 2 octopuses.

2 2 2 2 2 2

Two Two

2 2 2

PRACTICE HERE

2 2 2 2 2 2

PRACTICE HERE

2 2 2 2 2 2

I see 3 dolphins.

3 3 3 3 3 3

Three Three

3 3 3

PRACTICE HERE

3 3 3 3 3 3

PRACTICE HERE

3 3 3 3 3 3

I see 4 turtles.

4 4 4 4 4 4

Four Four

4 4 4

PRACTICE HERE

4 4 4 4 4 4

PRACTICE HERE

4 4 4 4 4 4

I see 5 squid.

5 5 5 5 5 5

Five Five

5 5 5

PRACTICE HERE

5 5 5 5 5 5

PRACTICE HERE

5 5 5 5 5 5

I see 6 clams.

6 6 6 6 6 6

Six Six

1 6 2 1 6 2 1 6 2

PRACTICE HERE

6 6 6 6 6 6

PRACTICE HERE

6 6 6 6 6 6

I see 7 oysters.

7 7 7 7 7 7

Seven Seven

1 → 7 2 ↓   1 → 7 2 ↓   1 → 7 2 ↓

PRACTICE HERE

7

PRACTICE HERE

7

I see 8 lobsters.

8 8 8 8 8 8

Eight Eight

8 8 8

PRACTICE HERE

8

PRACTICE HERE

8 8 8 8 8 8

I see 9 crabs.

9 9 9 9 9 9

Nine Nine

9 9 9

PRACTICE HERE

q q q q q q

PRACTICE HERE

q q q q q q

I see 10 shrimp.

10 10 10 10 10

Ten Ten

1 2 1 10 1 2 1 10 1 2 1 10

PRACTICE HERE

10 10 10 10 10

PRACTICE HERE

10 10 10 10 10

I see 11 jellyfish.

11 11 11 11

Eleven Eleven

11 11 11 11 11 11

PRACTICE HERE

PRACTICE HERE

11

I see 12 seahorses.

12 12 12 12 12

Twelve Twelve

12 12 12

PRACTICE HERE

12 12 12 12 12

PRACTICE HERE

12 12 12 12 12

I see 13 shells.

13 13 13 13 13

Thirteen Thirteen

13 13 13

PRACTICE HERE

13 13 13 13 13

PRACTICE HERE

13 13 13 13 13

I see 14 cuttlefish.

14 14 14 14 14

fourteen

14 14 14

PRACTICE HERE

14

PRACTICE HERE

14 14 14 14 14

I see 15 Polar Bears.

15 15 15 15 15

Fifteen

15 15 15

PRACTICE HERE

15 15 15 15 15

PRACTICE HERE

15 15 15 15 15

I see 16 seals.

16 16 16 16 16

Sixteen

16 16 16

PRACTICE HERE

16 16 16 16 16

PRACTICE HERE

16 16 16 16 16

I see 17 sharks.

17 17 17 17 17

Seventeen

17 17 17

PRACTICE HERE

17 17 17 17 17

PRACTICE HERE

17 17 17 17 17

I see 18 otters.

18 18 18 18 18

Eighteen

18 18 18

PRACTICE HERE

18 18 18 18 18

PRACTICE HERE

18 18 18 18 18

I see 19 walruses.

19 19 19 19 19

Nineteen

19 19 19

PRACTICE HERE

19 19 19 19 19

PRACTICE HERE

19 19 19 19 19

I see 20 penguins.

20 20 20 20

Twenty

20 20 20

PRACTICE HERE

20 20 20 20

PRACTICE HERE

20 20 20 20

I see 21 whales.

21 21 21 21 21

Twenty One

21 21 21

PRACTICE HERE

21 21 21 21 21

PRACTICE HERE

21 21 21 21 21

I see 22 dolphins.

22 22 22 22

Twenty Two

22 22 22

PRACTICE HERE

22 22 22 22

PRACTICE HERE

22 22 22 22

I see 23 clown fish.

23 23 23 23

Twenty Three

23 23 23

PRACTICE HERE

23 23 23 23

PRACTICE HERE

23 23 23 23

I see 24 pelicans.

24 24 24 24

Twenty Four

24 24 24

PRACTICE HERE

24 24 24 24

PRACTICE HERE

24 24 24 24

I see 25 undulate rays.

25 25 25 25

Twenty five

25 25 25

PRACTICE HERE

25 25 25 25

PRACTICE HERE

25 25 25 25

I see 26 Mermaids.

26 26 26 26

Twenty six

26 26 26

PRACTICE HERE

26 26 26 26

PRACTICE HERE

26 26 26 26

I see 27 angelfish.

27 27 27 27

Twenty Seven

PRACTICE HERE

27 27 27 27

PRACTICE HERE

27 27 27 27

I see 28 Starfish.

28 28 28 28

Twenty Eight

28 28 28

PRACTICE HERE

28 28 28 28

PRACTICE HERE

28 28 28 28

I see 29 Treasure Chests.

29 29 29 29

Twenty Nine

29 29 29

PRACTICE HERE

29 29 29 29

PRACTICE HERE

29 29 29 29

I see 30 tuna.

30 30 30 30

Thirty

30 30 30

PRACTICE HERE

30 30 30 30

PRACTICE HERE

30 30 30 30

I see 31 balloonfish.

31 31 31 31 31

Thirty One

31 31 31

PRACTICE HERE

31 31 31 31 31

PRACTICE HERE

31 31 31 31 31

I see 32 salmon.

32 32 32 32

Thirty-Two

32 32 32

PRACTICE HERE

32 32 32 32

PRACTICE HERE

32 32 32 32

I see 33 sea snails.

33 33 33 33

Thirty Three

33 33 33

PRACTICE HERE

33 33 33 33

PRACTICE HERE

33 33 33 33

I see 34 albacore.

34 34 34 34

Thirty Four

34 34 34

PRACTICE HERE

34 34 34 34

PRACTICE HERE

34 34 34 34

I see 35 sea lions.

35 35 35 35

Thirty Five

35 35 35

PRACTICE HERE

35 35 35 35

PRACTICE HERE

35 35 35 35

I see 36 sea nettles.

36 36 36 36

Thirty Six

36 36 36

PRACTICE HERE

36 36 36 36

PRACTICE HERE

36 36 36 36

I see 37 anchovy.

37 37 37 37

Thirty Seven

37 37 37

PRACTICE HERE

37 37 37 37

PRACTICE HERE

37 37 37 37

I see 38 dugong.

38 38 38 38

Thirty Eight

38 38 38

PRACTICE HERE

38 38 38 38

PRACTICE HERE

38 38 38 38

I see 39 cod.

39 39 39 39

Thirty Nine

39 39 39

PRACTICE HERE

39 39 39 39

PRACTICE HERE

39 39 39 39

I see 40 Royal Gramma Fish.

40 40 40 40

Forty

40 40 40

PRACTICE HERE

40 40 40 40

PRACTICE HERE

40 40 40 40

I see 41 clownfish.

41 41 41 41

Forty One

41 41 41

PRACTICE HERE

41

PRACTICE HERE

41

I see 42 sea urchins.

42 42 42 42

Forty Two

42 42 42

PRACTICE HERE

42 42 42 42

PRACTICE HERE

42 42 42 42

I see 43 Jellyfish.

43 43 43 43

Forty Three

43 43 43

PRACTICE HERE

43 43 43 43

PRACTICE HERE

43 43 43 43

I see 44 bull sharks.

44 44 44 44

Forty Four

PRACTICE HERE

44 44 44 44

PRACTICE HERE

44 44 44 44

I see 45 blue crabs.

45 45 45 45

Forty-Five

PRACTICE HERE

45 45 45 45

PRACTICE HERE

45 45 45 45

I see 46 blue whales.

46 46 46 46

Forty Six

46 46 46

PRACTICE HERE

46 46 46 46

PRACTICE HERE

46 46 46 46

I see 47 manta rays.

47 47 47 47

Forty Seven

47 47 47

PRACTICE HERE

47 47 47 47

PRACTICE HERE

47 47 47 47

I see 48 barracuda.

48 48 48 48

Forty Eight

48 48 48

PRACTICE HERE

48 48 48 48

PRACTICE HERE

48 48 48 48

I see 49 killer whales.

49 49 49 49

Forty Nine

49 49 49

PRACTICE HERE

49 49 49 49

PRACTICE HERE

49 49 49 49

I see 50 Mermaids.

50 50 50 50

Fifty

50 50 50

PRACTICE HERE

50 50 50 50

PRACTICE HERE

50 50 50 50

MATH ACTIVITIES

MATCH THE RIGHT ANSWER

3 

5 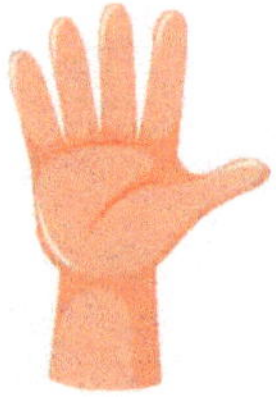

2 

4 

1 

SHADOW MATCH

DRAW A LINE TO THE MATCHING ANSWER.

MATCH THE RIGHT ANSWER...

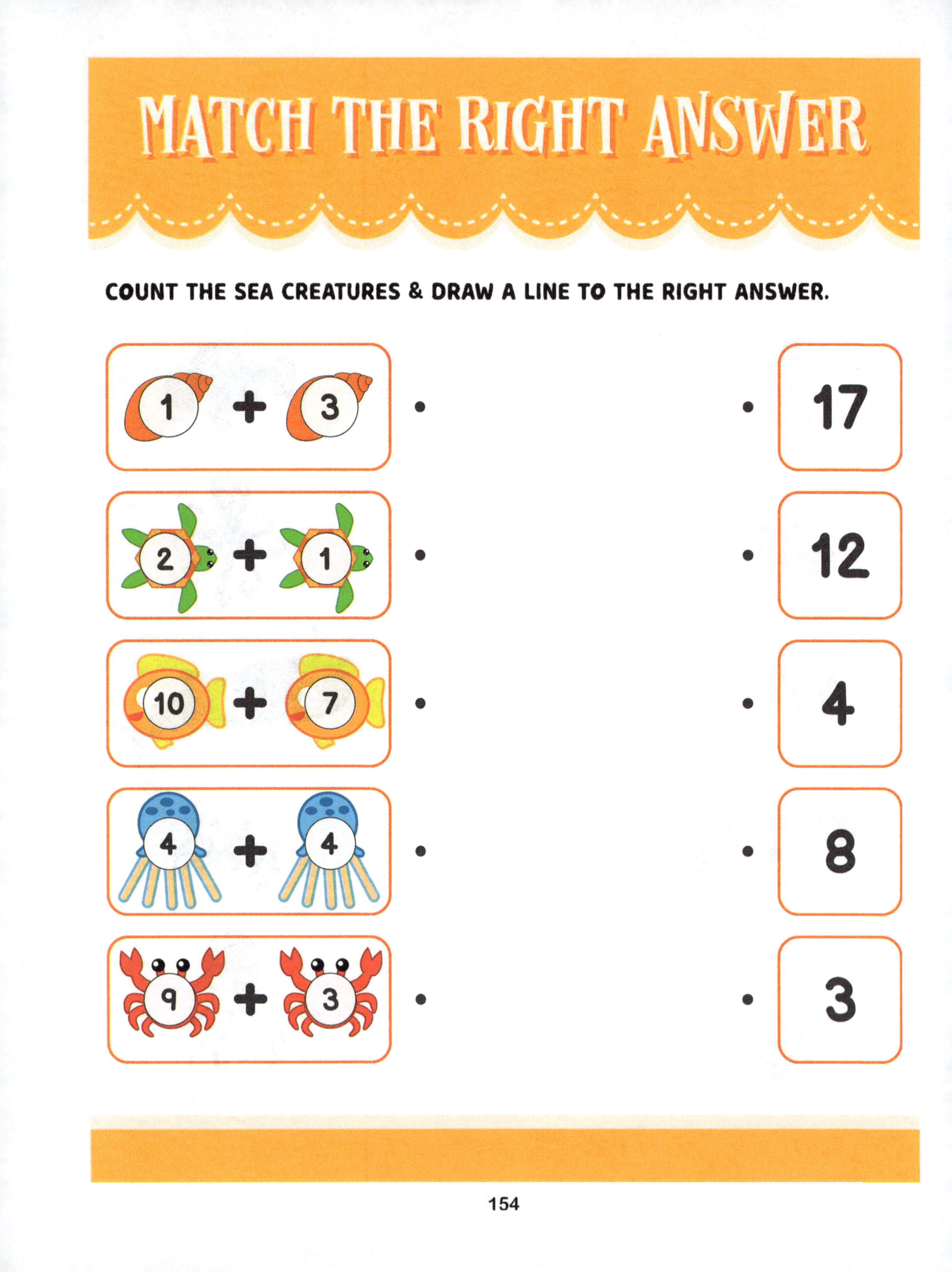

MATCH THE RIGHT ANSWER

COUNT THE SEA CREATURES & DRAW A LINE TO THE RIGHT ANSWER.

1 + 3	17
2 + 1	12
10 + 7	4
4 + 4	8
9 + 3	3

MATCH THE RIGHT SHADOW

DRAW A LINE TO THE MATCHING ANSWER.

MATCH THE RIGHT ANSWER

COUNT THE ITEMS & MATCH THE RIGHT ANSWER.

5

2

4

1

3

6

COUNTING PRACTICE

COUNT THE FISH & CIRCLE THE CORRECT NUMBER.

4 3 6

9 5 7

2 1 8

Counting is Fun

How many objects are in each box. Circle your answer and color the pictures

9 or 10

4 or 5

2 or 3

5 or 6

Counting is Fun

How many objects are in each box. Circle your answer and color the pictures

8 or 9

3 or 4

5 or 6

4 or 5

COUNT ALL THE SEA CREATURES

COUNT THE ITEMS & CIRCLE THE NUMBER.

1
2
4
3

3
4
5
2

6
7
9
8

5
7
6
9

WRITE THE MISSING NUMBER.

1		3	4	

COUNTING PRACTICE

COUNT THE ITEMS IN EACH SET. WRITE THE NUMBER.

TRACING PRACTICE

TRACE THE NUMBER WORD

1 One one one

2 Two two two

3 Three three

4 Four four four

5 Five five five

COUNT THE CRABS AND CIRCLE THE NUMBER WORD.

THREE	FIVE	TWO	FOUR	ONE

COMPARING NUMBERS GAME

GREATER THAN >   LESS THAN <   EQUAL TO =

< = >

FISH POND

TRACE THE DOTTED LINES & COLOR TO COMPLETE THE FISH PATTERN.

ADDITION

 +  = ☐

 + 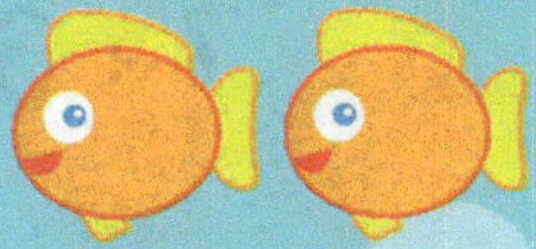 = ☐

 +  = ☐

 +  = ☐

 +  = ☐

Ocean Addition

3 + 3 = ____

7 + 3 = ____

4 + 4 = ____

4 + 6 = ____

2 + 2 = ____

5 + 1 = ____

2 + 6 = ____

1 + 3 = ____

8 + 2 = ____

Ocean Addition

Solve the problems and color the pictures

3 + 2 = ____

2 + 4 = ____

1 + 5 = ____

4 + 5 = ____

3 + 2 = ____

4 + 1 = ____

6 + 3 = ____

4 + 2 = ____

2 + 2 = ____

Ocean Addition

Solve the problems and color the pictures

4 + 3 = ____

4 + 5 = ____

2 + 6 = ____

5 + 6 = ____

4 + 4 = ____

5 + 2 = ____

7 + 4 = ____

5 + 3 = ____

3 + 3 = ____

Ocean Addition

Solve the problems and color the pictures

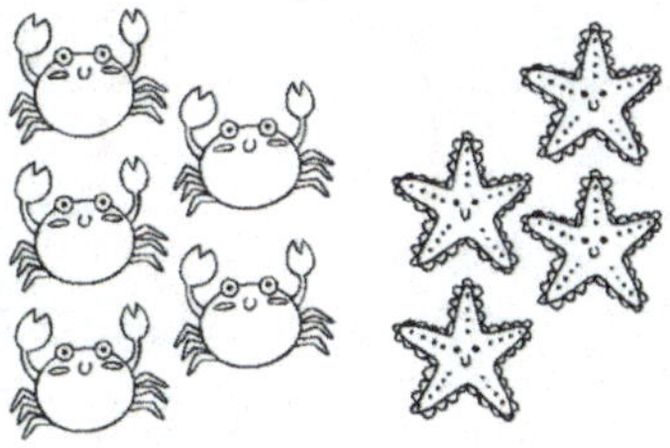

5 + 4 = ___

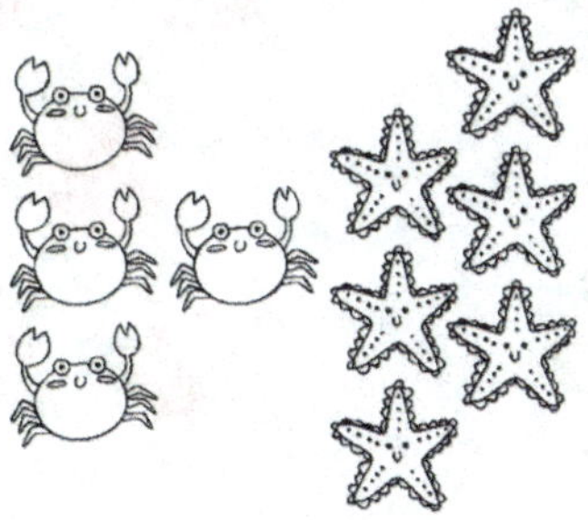

4 + 6 = ___

3 + 7 = ___

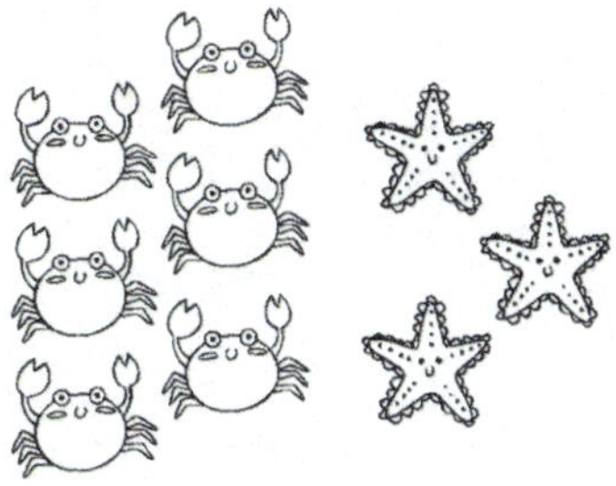

6 + 3 = ___

1 + 5 = ___

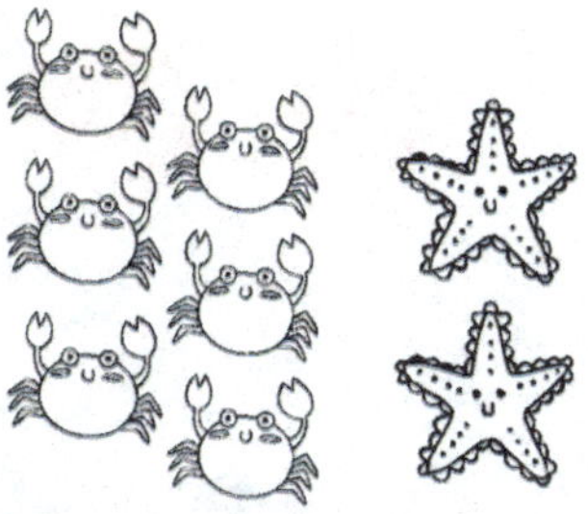

6 + 2 = ___

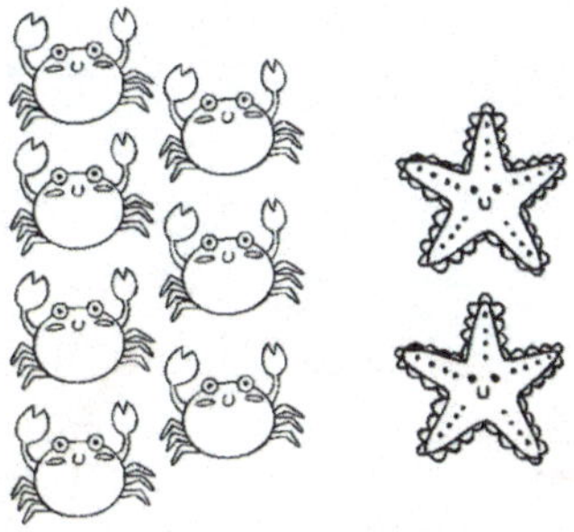

7 + 2 = ___

4 + 1 = ___

3 + 2 = ___

SUBTRACTION

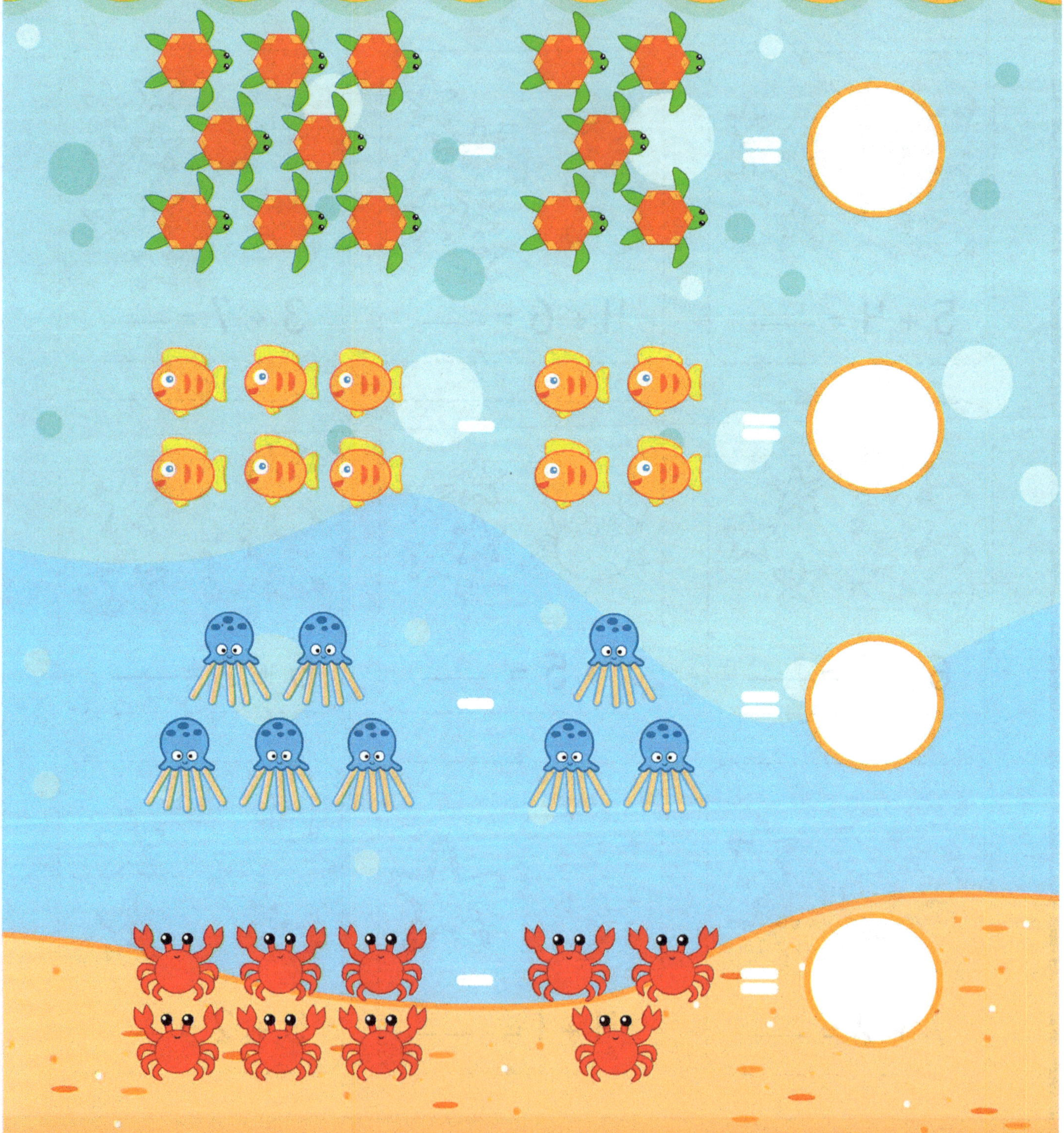

Ocean Subtraction

5 - 2 = ____

3 - 2 = ____

4 - 1 = ____

5 - 5 = ____

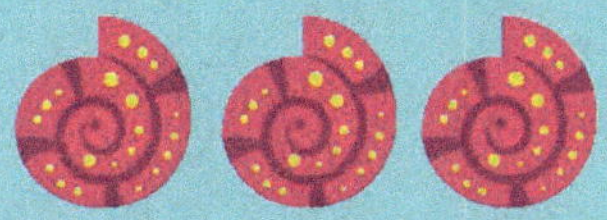

3 - 1 = ____

4 - 3 = ____

4 - 2 = ____

2 - 1 = ____

Ocean Subtraction

Solve the problems and color the pictures

4 - 2 = ___

2 - 2 = ___

5 - 1 = ___

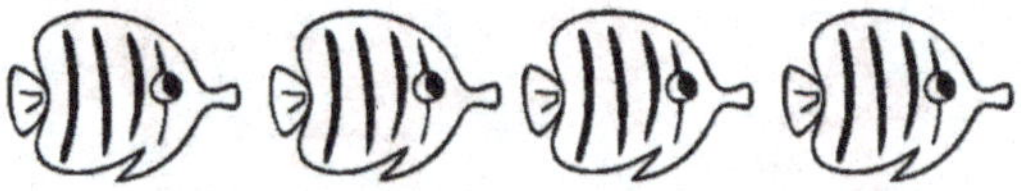

4 - 3 = ___

3 - 2 = ___

5 - 3 = ___

4 - 1 = ___

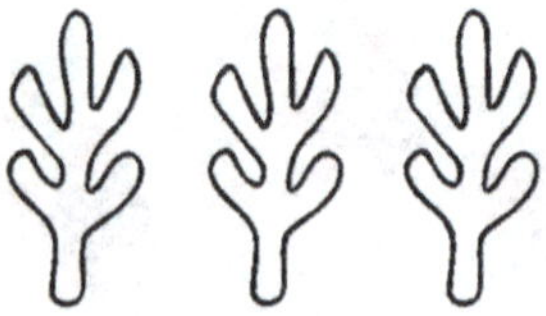

3 - 1 = ___

Ocean Subtraction

Solve the problems and color the pictures

4 - 2 = ____

3 - 1 = ____

6 - 4 = ____

8 - 5 = ____

7 - 4 = ____

5 - 3 = ____

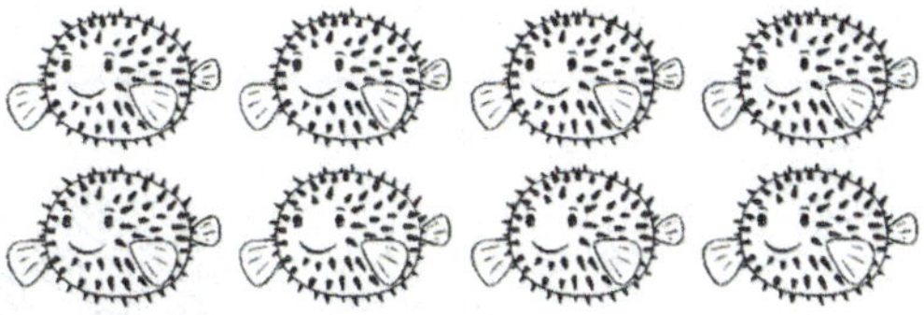

8 - 4 = ____

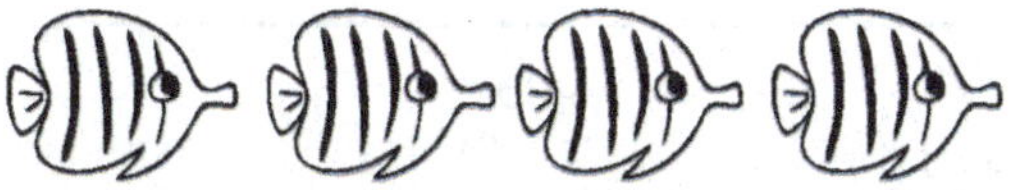

4 - 1 = ____

Ocean Subtraction

Solve the problems and color the pictures

4 - 2 = ___

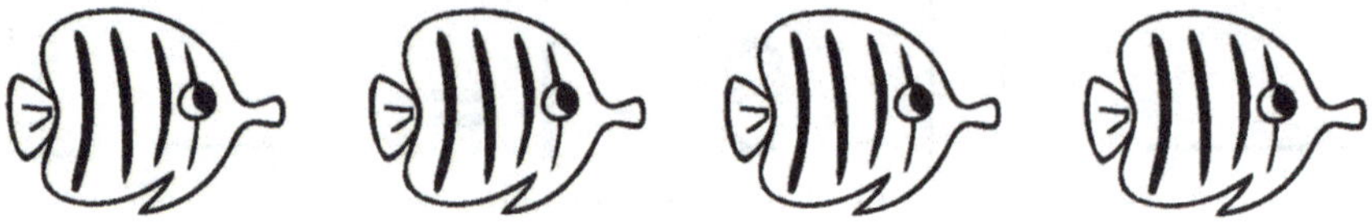

4 - 3 = ___

7 - 1 = ___

3 - 2 = ___

8 - 5 = ___

10 - 7 = ___

HOW MANY OCTOPUSES?

GREEN ☐ RED ☐ ORANGE ☐

PURPLE ☐ PINK ☐

How Many Are There?

COUNT THE SEA CREATURES & CIRCLE THE ANSWER

4 2 6

4 5 1

6 1 3

3 5 2

2 6 7

5 8 9

WRITE THE MISSING NUMBER.

	2		4	5

How Many Are There?

COLOR BY NUMBER

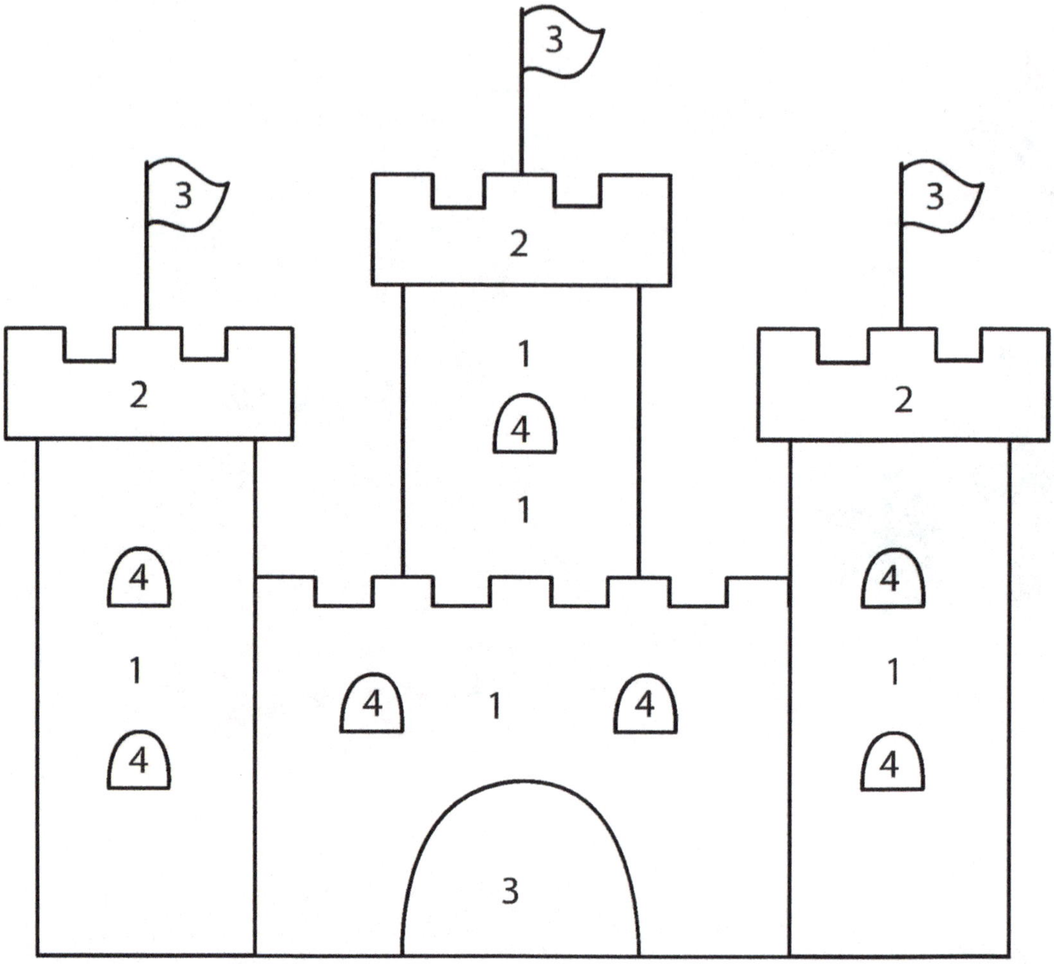

1 -

2 -

3 -

4 -

Under The sea

Color and Count

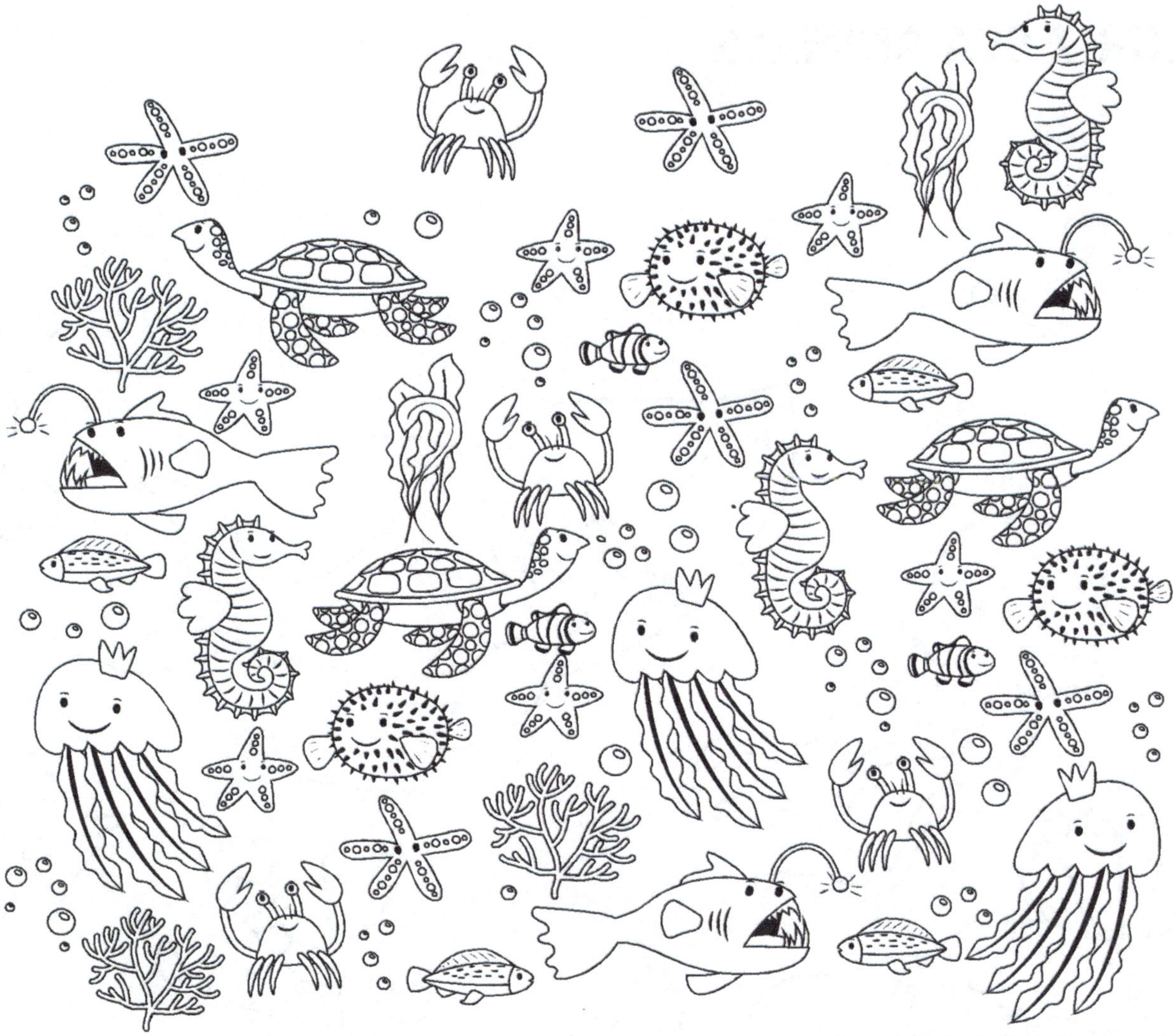

How many?

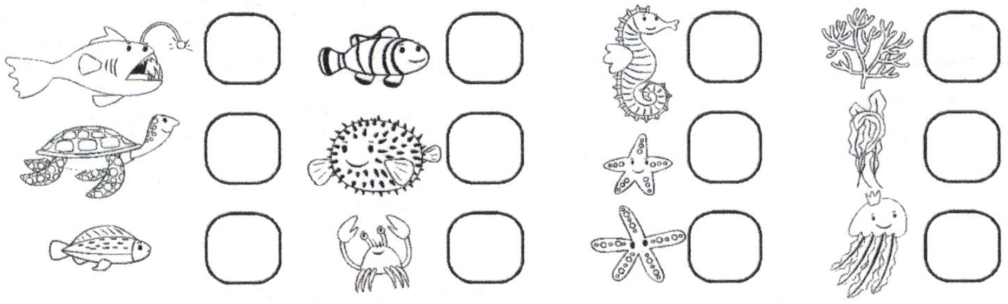

COLOR IT

COLOR 7 SEA ANIMALS

DRAW & COLOR

DRAW 5 SPOTS ON THE DOLPHIN AND COLOR THEM.

FIND THE NUMBER

CIRCLE EVERY NUMBER 2.

SPOT THE DIFFERENCE

WRITE THE MISSING NUMBER.

5		7		9

TOP & BOTTOM

COLOR THE ANIMAL ON THE TOP

COLOR THE ANIMAL ON THE BOTTOM

COLOR THE ANIMAL ON THE BOTTOM

COLOR THE ANIMAL ON THE TOP

LEFT & RIGHT

COLOR THE ANIMAL ON THE LEFT

COLOR THE ANIMAL ON THE RIGHT

CUT & PASTE

CUTTING

1 4 3 2 6 5

CUT & PASTE

CUTTING

JELLYFISH	CRAB	SHARK
WHALE	OCTOPUS	CLOWNFISH

Under The sea

Trace the dotted lines to complete the picture.

Ocean Pencil Control

Trace along the dotted line to guide the sea animals to the ocean

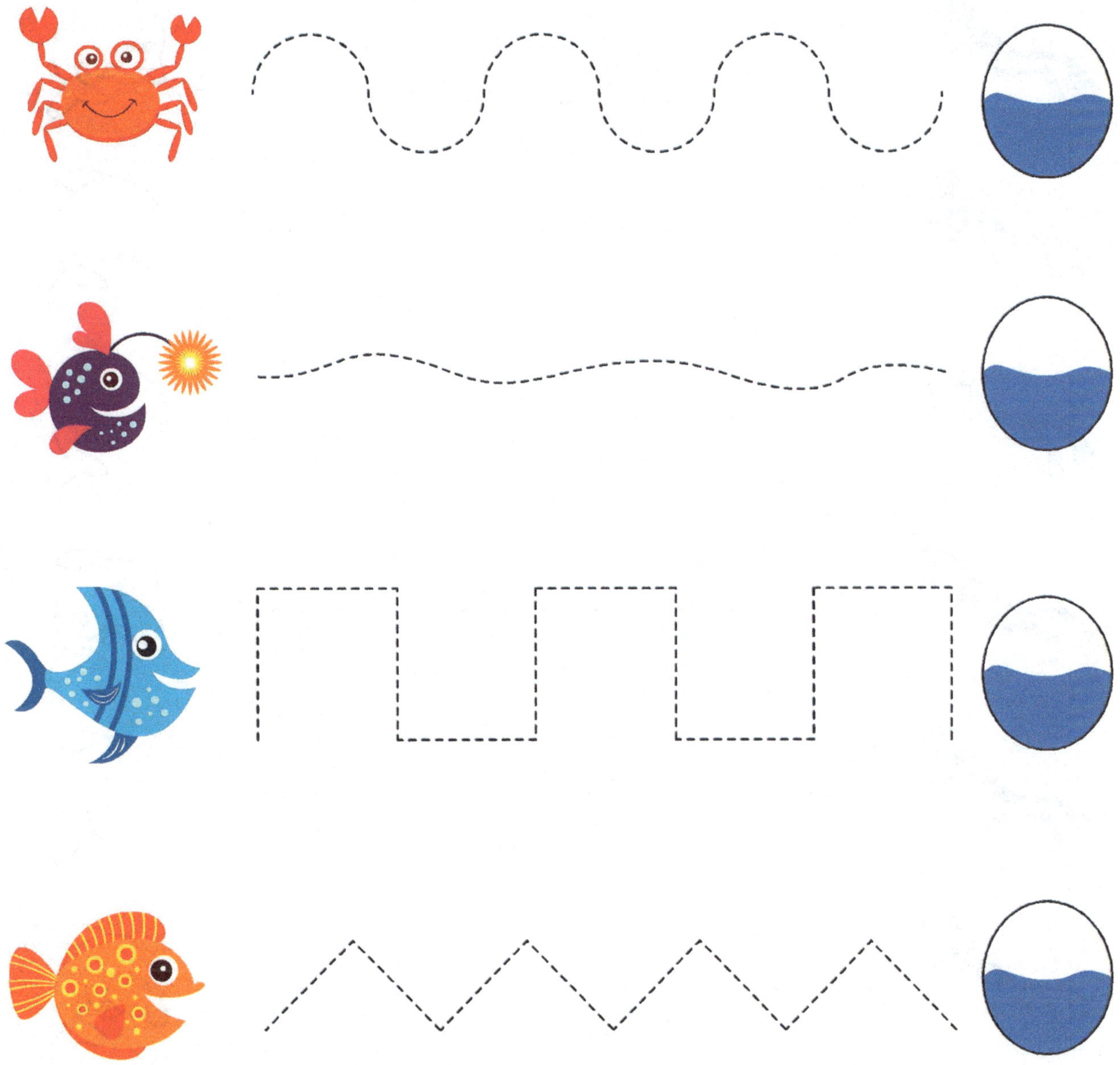

Ocean Pencil Control

Trace the lines to help the diver find the Dolphin. Then color the Dolphin

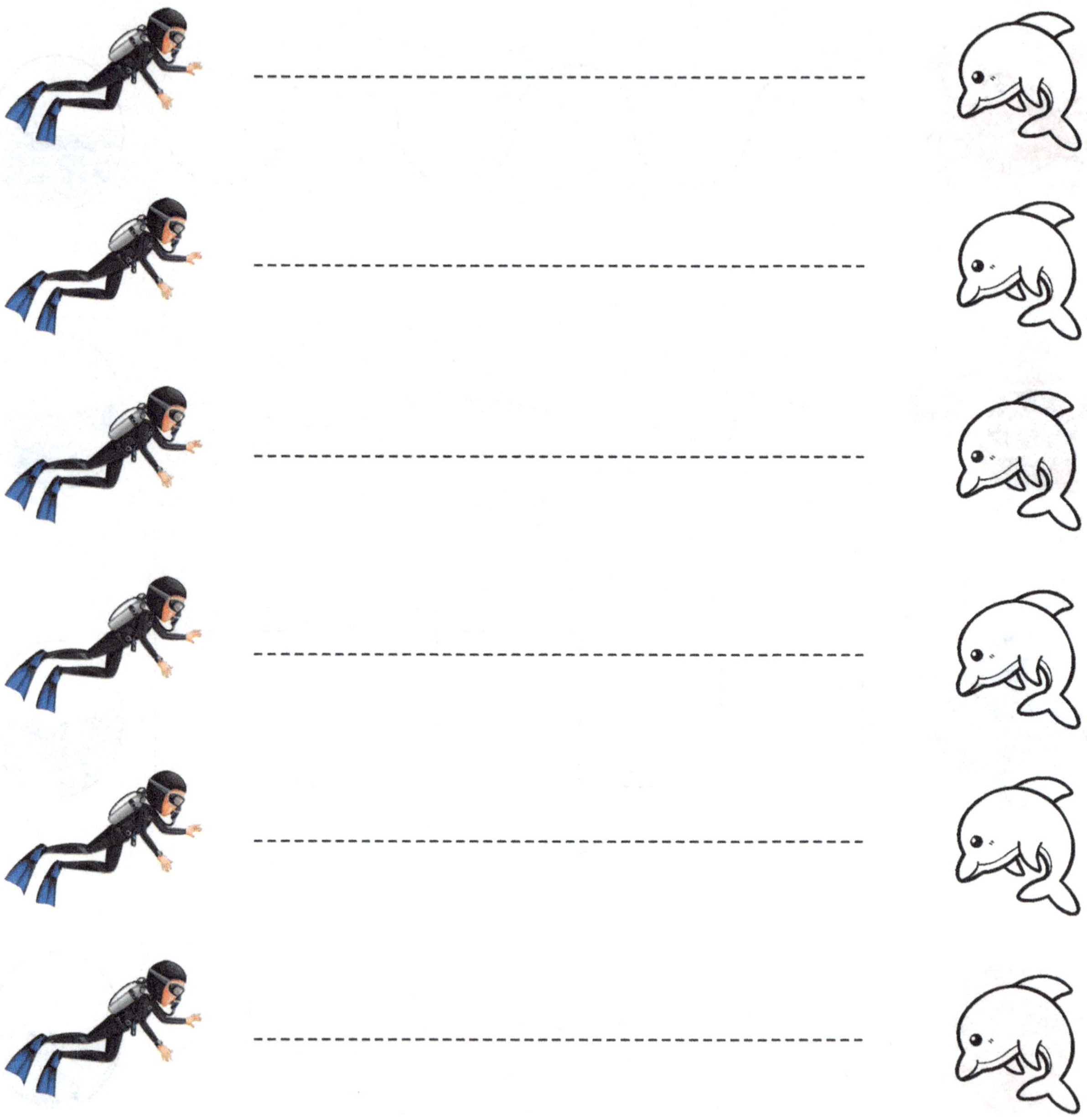

Ocean Pencil Control

Trace the lines to help the diver find the Seahorse. Then color the Seahorse

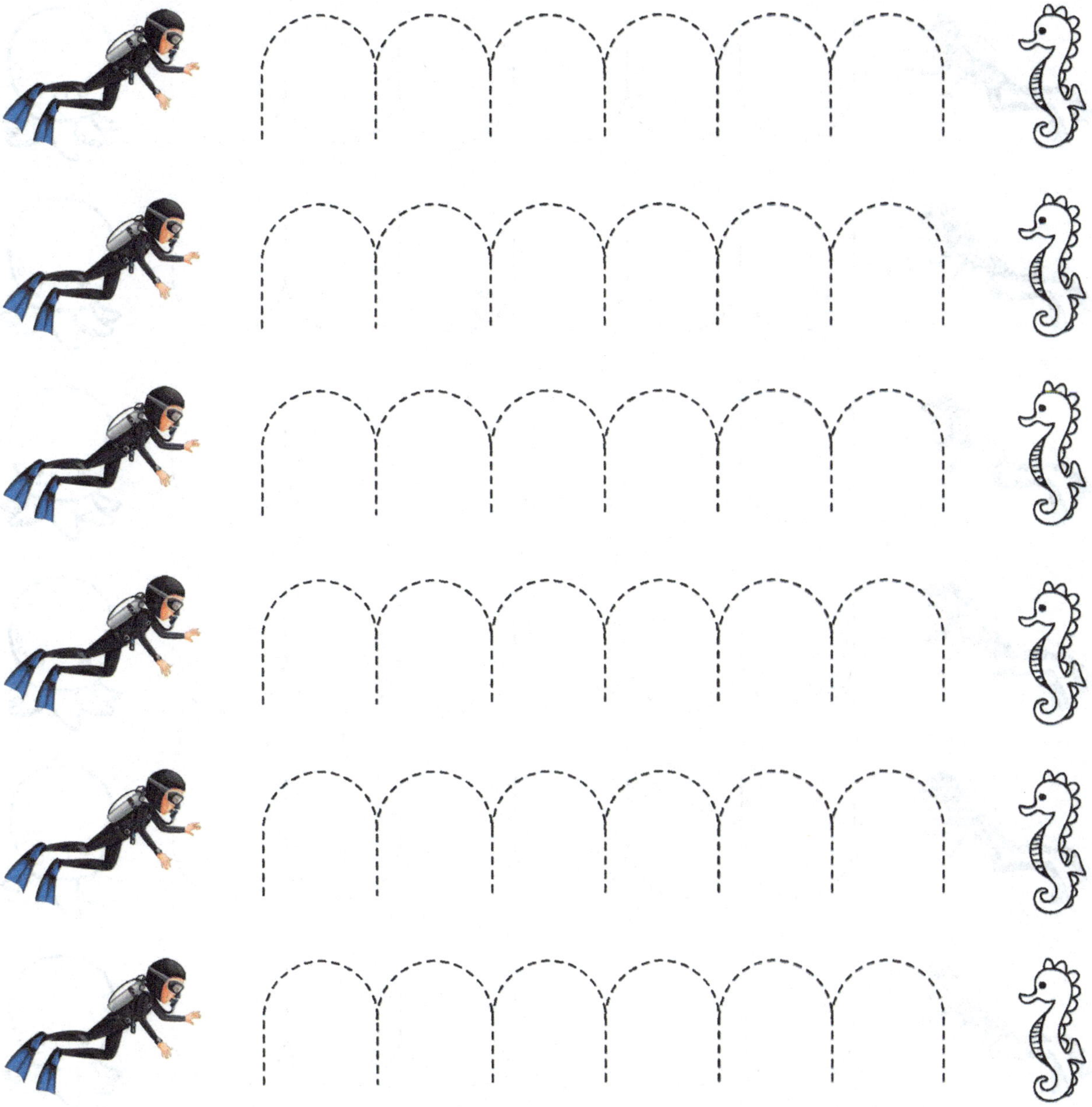

Ocean Pencil Control

Trace the lines to help the diver find the Octopus. Then color the Octopus

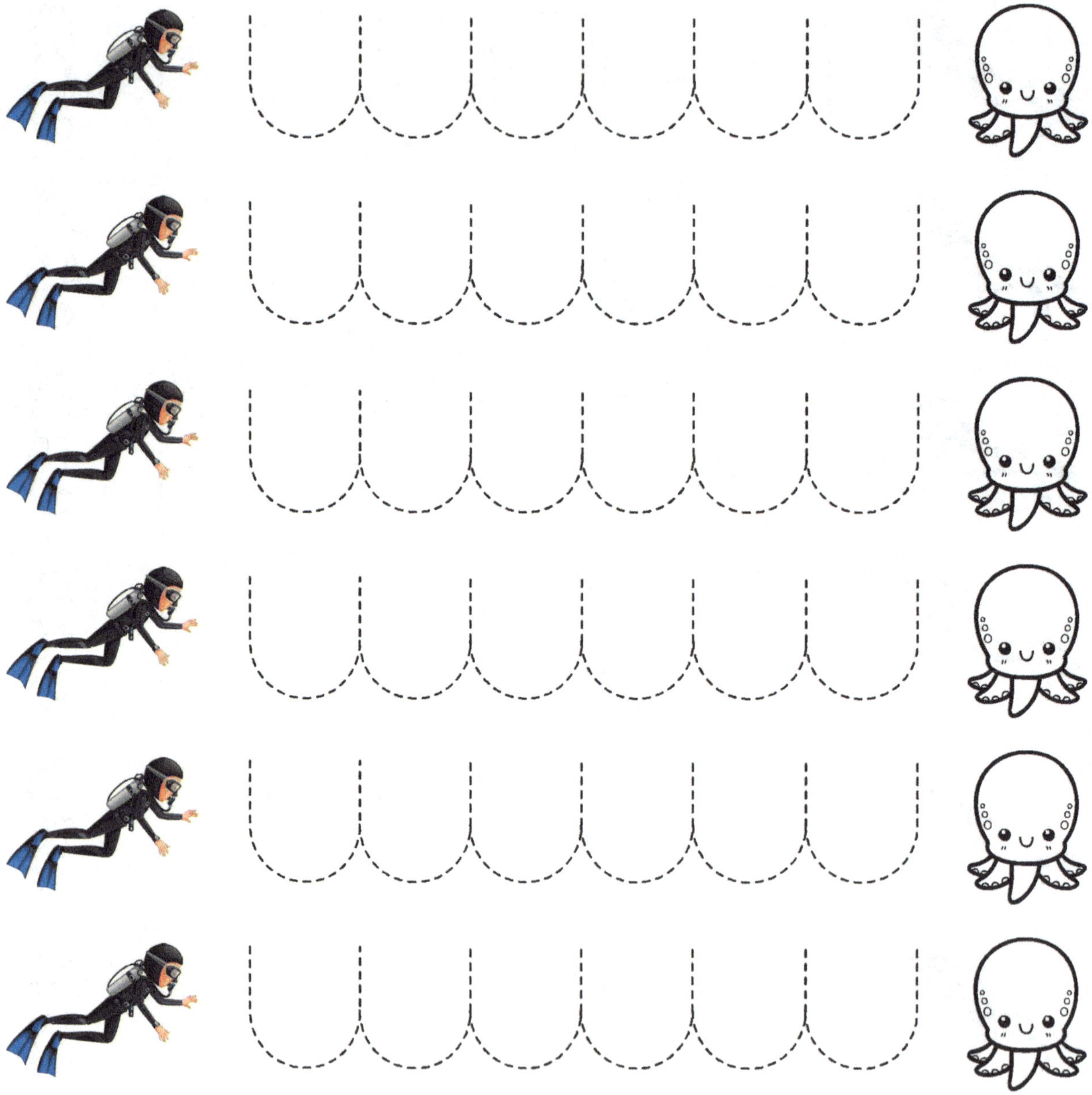

Ocean Pencil Control

Trace the lines to help the diver find the Crab. Then color the Crab.

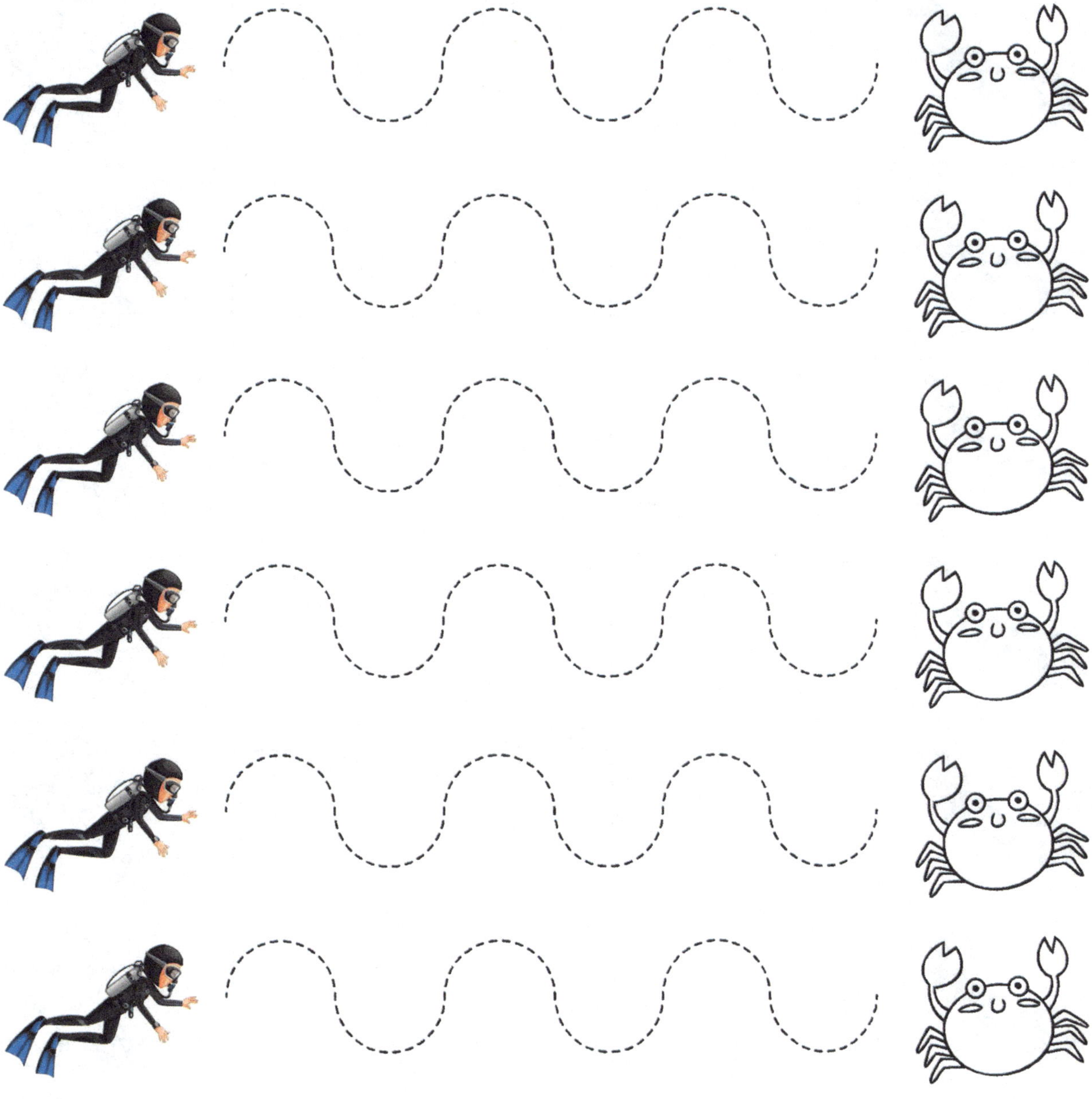

Ocean Pencil Control

Trace the lines to help the diver find the Starfish. Then color the Starfish

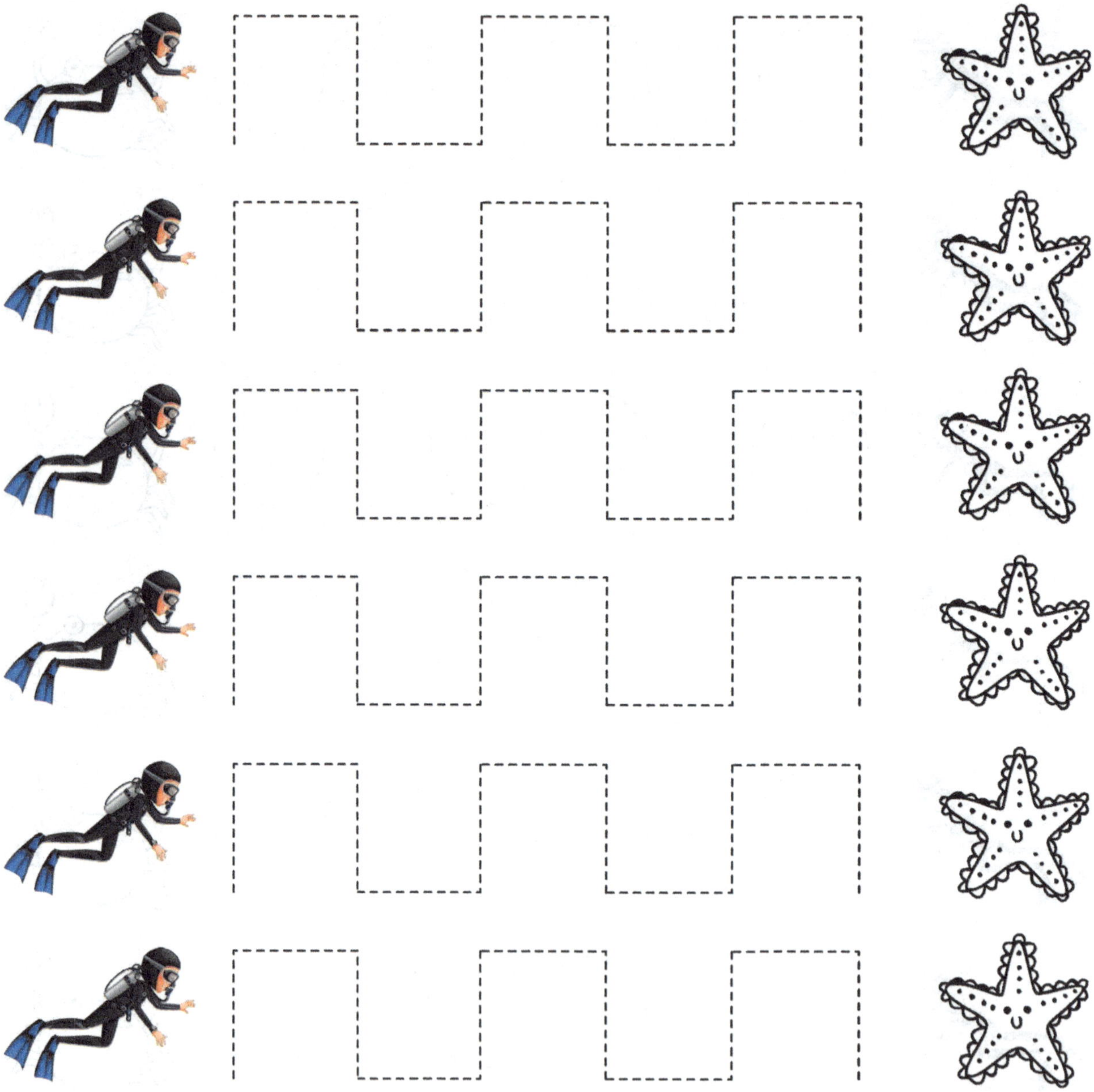

Ocean Pencil Control

Trace the lines to help the diver find the Sea Turtle. Then color the Sea Turtle

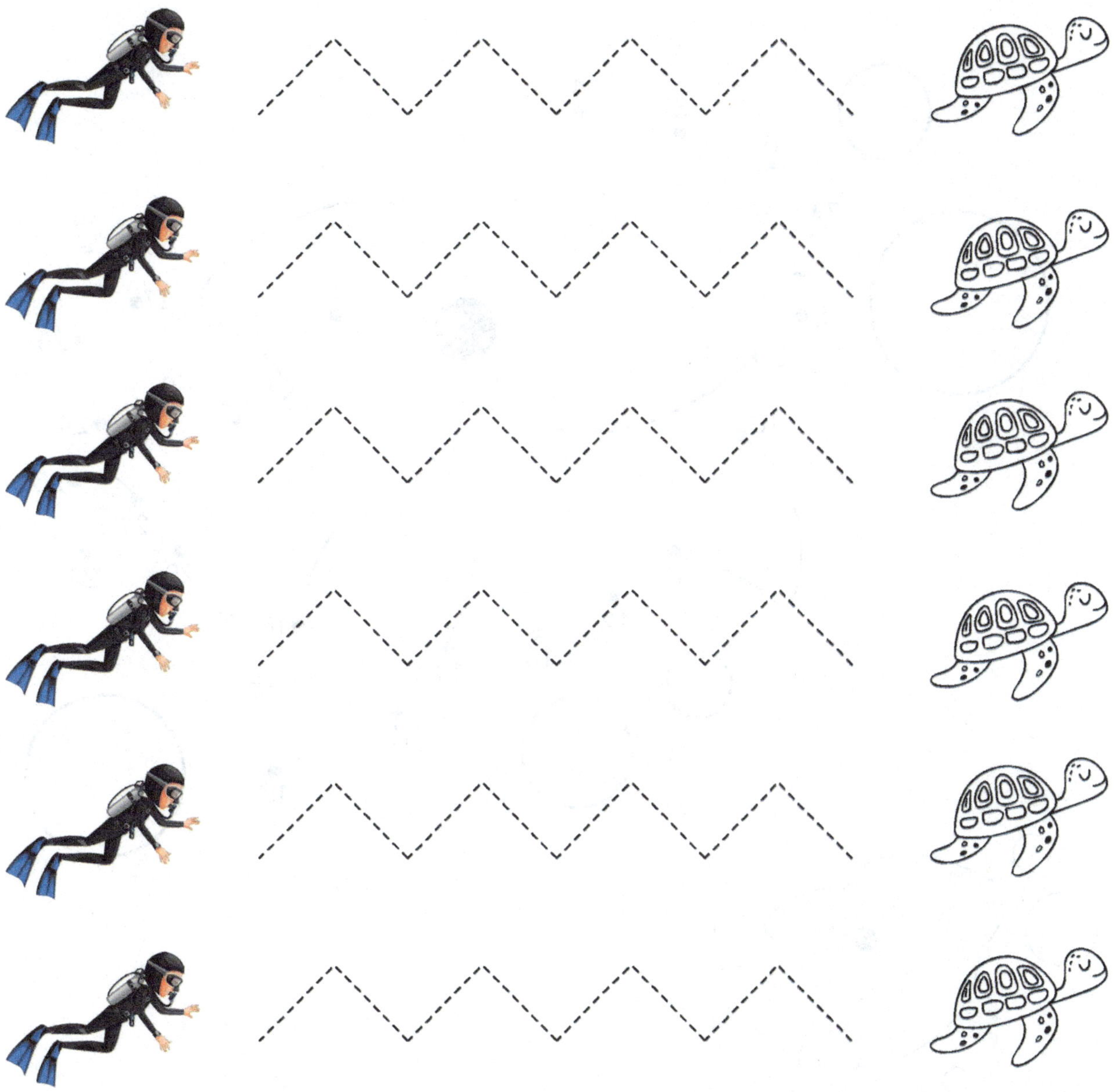

Dot to Dot

Join the dots to complete the sea creature and color it.

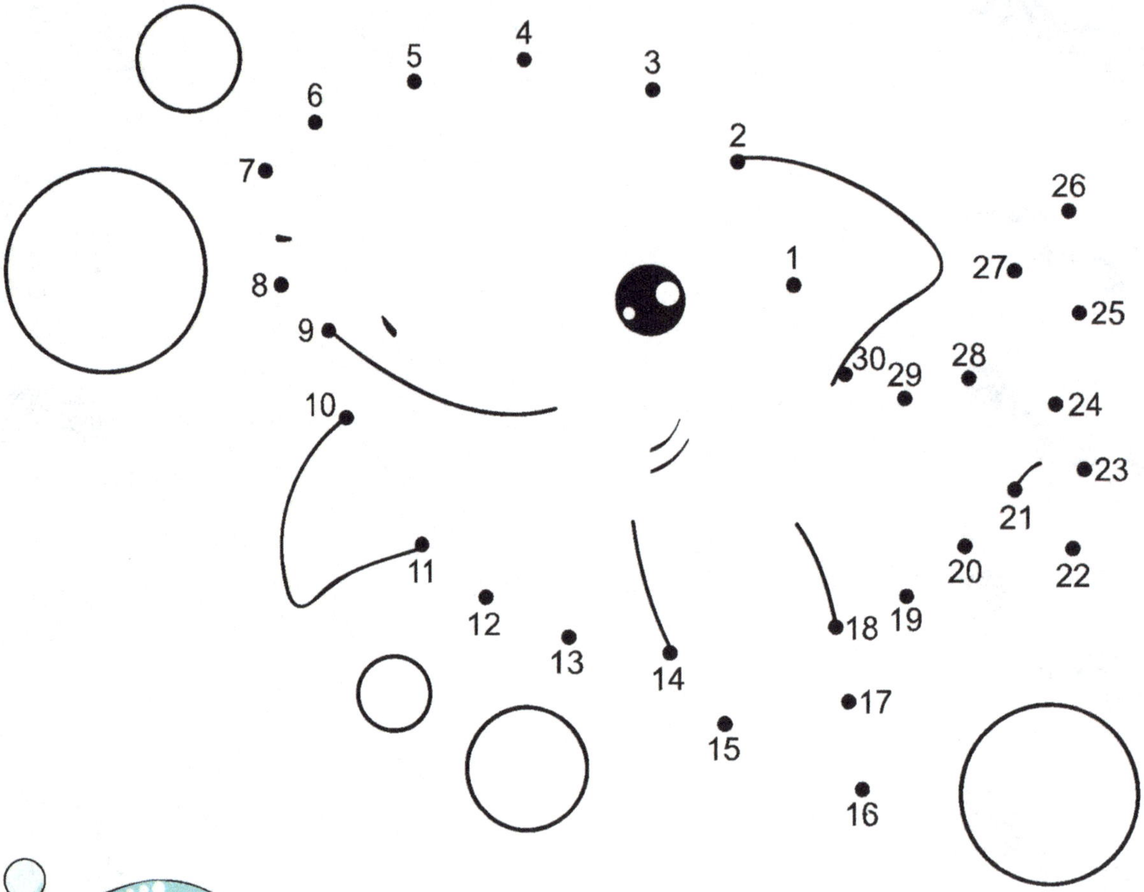

Dot to Dot

Join the dots to complete the sea creature and color it.

Dot to Dot

Join the dots to complete the sea creature and color it.

Dot to Dot

Join the dots to complete the sea creature and color it.

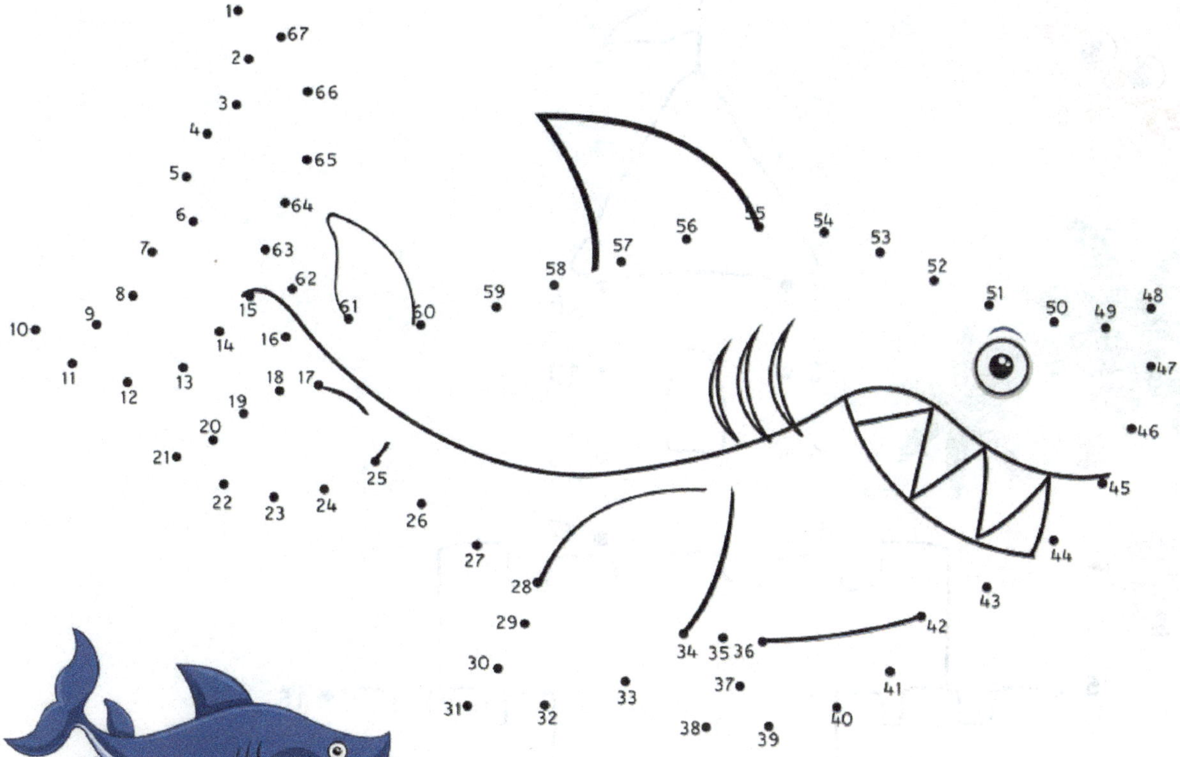

SANDCASTLE

JOIN THE DOTS AND COLOR THE SANDCASTLE.

MAZE GAME

A WINDING WAY

HELP THE PIRATE FIND TREASURE BY TAKING THE CORRECT PATH.

TURTLE PATH

HELP THE TURTLE REACH THE FINISHING BOARD BY DRAWING A LINE.

HELP THE GIRL GET TO THE SEA

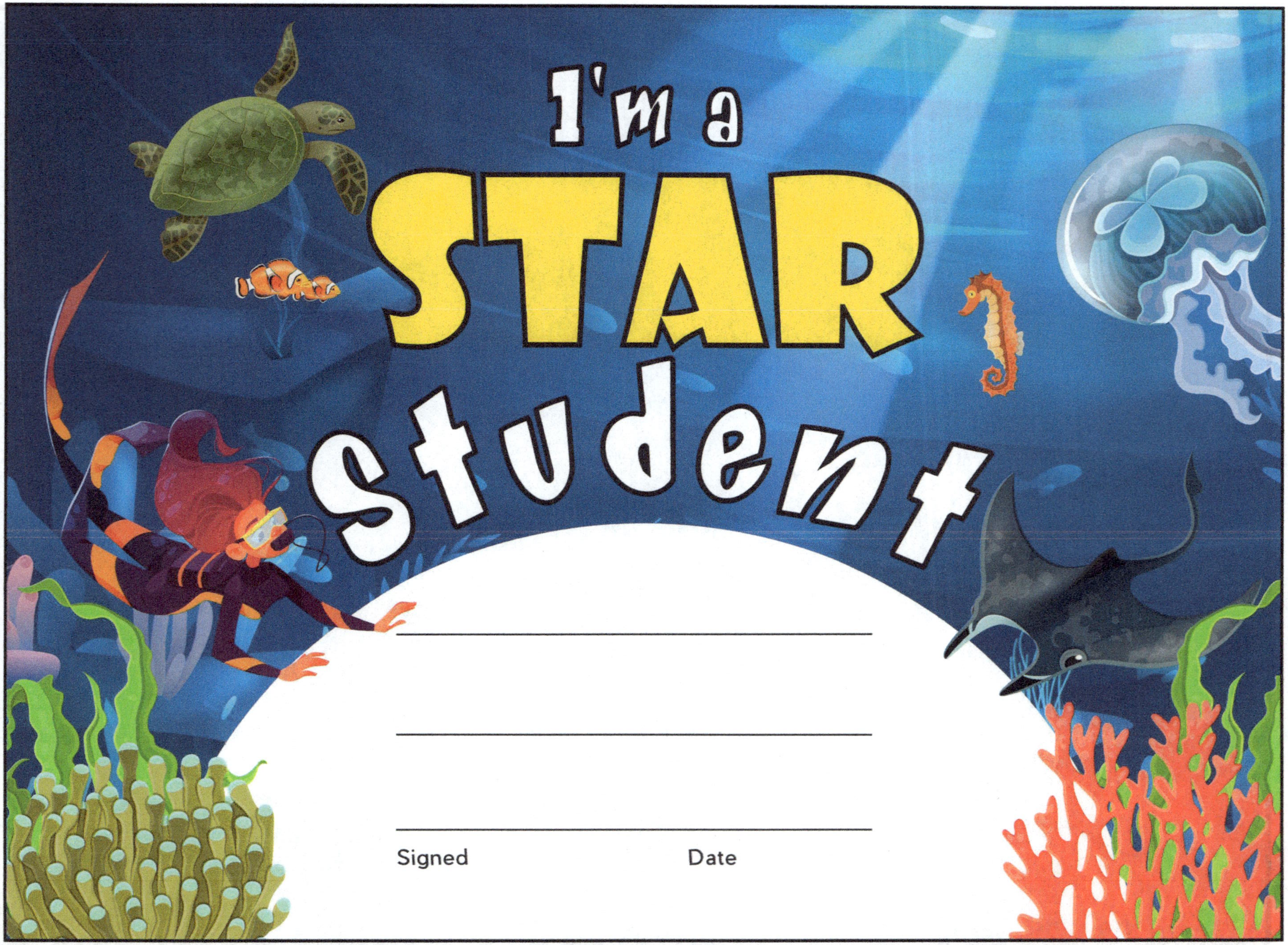

I'm a
STAR
Student
Signed
Date

Made in the USA
Monee, IL
02 November 2026

67376093R00116